Deivid Nogueira Rafael
Simone Aquino

Electronic Prescription

AF376722

Deivid Nogueira Rafael
Simone Aquino

Electronic Prescription

ScienciaScripts

Imprint
Any brand names and product names mentioned in this book are subject to trademark, brand or patent protection and are trademarks or registered trademarks of their respective holders. The use of brand names, product names, common names, trade names, product descriptions etc. even without a particular marking in this work is in no way to be construed to mean that such names may be regarded as unrestricted in respect of trademark and brand protection legislation and could thus be used by anyone.

Cover image: www.ingimage.com

This book is a translation from the original published under ISBN 978-620-2-18720-6.

Publisher:
Sciencia Scripts
is a trademark of
Dodo Books Indian Ocean Ltd. and OmniScriptum S.R.L publishing group

120 High Road, East Finchley, London, N2 9ED, United Kingdom
Str. Armeneasca 28/1, office 1, Chisinau MD-2012, Republic of Moldova, Europe
Printed at: see last page
ISBN: 978-620-7-26455-1

Copyright © Deivid Nogueira Rafael, Simone Aquino
Copyright © 2024 Dodo Books Indian Ocean Ltd. and OmniScriptum S.R.L publishing group

1.1. Medical Prescription: Concepts and Applications

The hospital prescription can be defined as the first stage in the supply of medicines to a hospitalised patient, being part of a process that involves various professionals such as doctors, pharmacists, nutritionists and nurses, and is extremely complex. It is the transition from therapeutic planning to the implementation of the support team in the patient's treatment (Pazin-Filho *et al.*, 2013), and is essentially a communication tool between doctor, pharmacist, nurse, carer and patient (Néri *et al.*, 2011). Prescribing medicines is also an important activity in the health care process, representing one of the medical actions possible in the patient's medical records (Wen, 2000).

The prescription is seen as the first stage in a series of events within the medication procedure, which will result in the safe or unsafe administration to the patient. In this case, all prescribing professionals need to be aware of the need for their prescriptions to be clear, objective and complete, minimising the doubts of all professionals and providing favourable conditions for drug therapy (Gimenes *et al.*, 2010).

According to the Resolution of the Federal Council of Medicine (CFM) n. 1.931/09, set out in Chapter III, article 11 of the Code of Medical Ethics, it is forbidden for a doctor to "Prescribe, certify or issue reports in a secret or illegible manner, without duly identifying their registration number with the Regional Council of Medicine in their jurisdiction". According to Ordinance 3916 of 30 October 1998, prescription is the act of defining the medicine to be taken by the patient, with the respective dosage and duration of treatment. In general, this act is expressed by drawing up a prescription. It is a systematised activity in interrelated stages, which provide important elements for a correct prescription. It is therefore essential that prescriptions have the minimum elements needed to achieve an efficient care process (Néri, Viana, & Campos, 2008). Prescriptions can be classified as described in Table 1:

Table 1. Classification of prescription types.

In terms of origin	Outpatient if it comes from an outpatient clinic; Hospital when it is carried out for an in-patient.
Types of Medicines Prescribed	Manipulated when the doctor selects the inputs, quantities and pharmaceutical form; With a pharmaceutical speciality when it contains a market drug, produced by the pharmaceutical industry and which must be administered in the form supplied, without alteration.
Types of Prescription	Urgent when it indicates the need for immediate treatment; PRN from the Latin "pro re nata", are those "if necessary", the prescribed treatment is administered according to the patient's needs, taking into account their condition and respecting the minimum time between administrations; Protocol-based when they have specific, pre-established criteria; This is the most common pattern, where treatment is started and continues until the doctor stops it; Standard with closing date, indicating the start and end of treatment.

Source: Adapted from Néri, Campos and Viana (2008).

According to Chapter VI, Article 35 of Law 5.991/1973, hospital prescriptions must be complete, legible and clear. In addition, it must contain the patient's name, registration number and bed, date, name of the medicine, dosage, route, frequency and time of administration, duration of treatment, the doctor's legible signature and registration number with the Regional Council of Medicine (CRM). In the case of outpatients, it is also necessary to provide the patient's home address and the prescriber's office or residence (Lopes *et al.*, 2014). Table 2 shows the basic elements contained in a prescription:

Table 2. Basic elements contained in a prescription.

Features	Considerations
Header	Name of the institution and its address.
Oversubscription	Patient data such as name, address, in this case floor, ward, service, bed and medical record number, age, weight, height, allergies.
Enrolment	Name of the medicine according to the DCB or DCI, concentration, using units of weights and measures of the national metric system and pharmaceutical form.
Sub-registration	Dose; It is expressed in units of weights and measures of the national metric system, diluent: type and volume, posology, total quantity to be dispensed and administered, route of administration, speed of infusion and duration of therapy.
Transcript	Composed of the prescriber's instructions to the pharmacist and/or nurse.
Date	Date.
Identification of the Prescriber	Stamp with registration number with the Regional Medical/Dental Council and signature.

Source: Adapted from Néri, Campos and Viana (2008) and Pazin-Filho *et al.* (2013).

In 1968 it was observed that by administering all nutrients only intravenously, a child could achieve normal development and growth (Dudrick, 1968 as quoted in Rocamora, 2001 p.291). This event led to enthusiasm for this more modern nutritional therapy, due to its ready availability as well as that of the main components that make up parenteral nutrition such as: amino acid hydrolysates, hypertonic dextroses, lipid emulsions, electrolytes and vitamins, as well as its ease of central venous catheterisation, thus spreading this type of nutrition all over the world (Waitzberg, 2002).

Nutritional Therapy (NT) is understood to be the set of therapeutic procedures aimed at maintaining or recovering nutritional status by means of Parenteral Nutrition (PN) or Enteral Nutrition (NE), carried out on patients unable to adequately meet their nutritional and metabolic needs orally (RDC n. 60, 2000). The objectives of NT include correcting previous malnutrition, preventing or attenuating the caloric-protein deficiency that usually occurs during the course of the illness that led to hospitalisation, balancing the metabolic state with the administration of fluids, nutrients and electrolytes and reducing morbidity with a consequent reduction in the recovery period (Carvalho *et al.*, 2014). The prevalence of malnutrition in hospitalised patients varies between 30% and 50%, according to studies carried out in different countries. This prevalence increases with the length of hospitalisation, affecting 61% of patients hospitalised for more than 15 days. There is also iatrogenic malnutrition due to negligence in providing nutritional support (Carvalho *et al.*, 2014).

To prepare PN formulas, professionals prescribe and use quantities of small volume parenteral solutions (SPPVs) such as vitamins, minerals and electrolytes, and large volume parenteral solutions (SPGVs) such as sterilised water, lipids, amino acid sources and dextrose (Ferreira, 2007). Parenteral nutrition is handled in a compounding pharmacy specialising in the preparation of personalised medicines for patients, where they are prepared on the basis of a prescription prescribed by medical professionals, the individual ingredients of which are mixed together in the form of an exact dosage. This personalised method allows the compounding pharmacist to work with a drug that meets the patient's specific needs (*Professional Compounding Centers of America, Inc.*, 2015).

PN is an apyrogenic and sterile formulation composed basically of amino acids, lipids, vitamins and minerals, and can be administered peripherally or centrally (Bottoni, Hassan, Nacarato, Garnes, & Bottoni, 2014). The Ministry of Health and Health Surveillance Ordinance No. 272 (1998) defines PN as:

> [...] solution or emulsion, basically composed of carbohydrates, amino acids, lipids, vitamins and minerals, sterile and apyrogenic, packaged in a glass or plastic container, intended for intravenous administration in malnourished or non-malnourished patients, in hospital, outpatient or home settings, aiming at the synthesis or maintenance of tissues, organs or systems.

It is prescribed when it is partially or totally impossible to use the gastrointestinal tract, i.e. it is indicated when the patient is unable to use the enteral route to meet their nutritional needs or in cases where they have underlying diseases that cause impairment to the ingestion, digestion or absorption of food (Rodrigues & Sobreira, 2013; Fletcher, 2013). It has been used in cancer patients as an enabling therapy for antineoplastic treatments such as chemotherapy and radiotherapy, improving nutritional status, minimising toxicities and prolonging patient survival (Garófolo, Boin,

Modesto, & Petrilli, 2007).

The handling of parenteral nutrition is dedicated and adapted to the individual needs of each patient, and its formulations are prescribed according to age, gender, the patient's condition and their specific illness (Ansel & Stoklosa, 2008). Its handling technique is considered a high-risk operation, as it involves numerous and complex aseptic manipulations carried out over a prolonged period. These preparations must be made from sterile components and with procedures that exclude the access of viable microorganisms, since the final solution cannot withstand terminal sterilisation (Mendonça & Silva, 2015). NP formulations are considered high-risk sterile preparations because they are complex aseptic manipulations in which a large number of substances are used during preparation. Controlling the work area, the aseptic technique, the handlers and the nutrients to be used to prepare parenteral nutrition allows the mixtures to result in pharmaceutical products suitable for administration (RDC n. 17, 2010). Maintaining the sterility of NP depends on the quality of the components incorporated and the previously validated environmental/technical conditions used during the handling procedure (Ordinance MS/SVS n. 272, 1998).

For NPs to result in suitable pharmaceutical products that can be administered safely, it is necessary to control the work area, the aseptic technique, the handlers and the parenteral nutrition solution being handled (Mendonça & Silva, 2015). According to Ordinance MS/SVS no. 272 (1998), it must be handled by a qualified professional following the correct dressing techniques, wearing sterilised clothing and sterile gloves after washing their hands, using the appropriate technique and antiseptic.

There are two types of PN, Total Parenteral Nutrition (TPN) and Partial Parenteral Nutrition (PPN). In TPN, all the essential nutrients are provided in adequate quantities for the total maintenance of life, and should include minerals, trace elements, vitamins, carbohydrates, lipids, amino acids and electrolytes. It is administered to the patient via a central catheter, allowing the administration of hyperosmolar solutions with minimal inconvenience. PPN is made up of low-osmolality solutions and is indicated for short-term nutritional maintenance, complementing oral intake and providing part of the daily nutritional needs (Ansel & Stoklosa, 2008).

Complications can occur in the administration of both PTN and PPN due to the methods of introducing and maintaining the catheter, leading to sepsis and treatment that involves removing the catheter. TPN is a process adapted to the needs of each patient, its formulations differ according to the age groups, gender and specific diseases of each patient, and its monitoring is essential for the prevention of metabolic and septic complications (Ansel & Stoklosa, 2008).

In Brazil, considering the importance of NT and due to the need to guarantee patients adequate nutritional care, the Ministry of Health, through the Health Surveillance Secretariat (SVS/ MS), through Ordinance 272/98, regulates nutritional therapy, with the formation of a Multiprofessional Nutritional Therapy Team (EMTN), establishing good parenteral nutrition preparation practices (BPPNP) and good parenteral nutrition administration practices (BPANP), thus regulating the stages of medical indication and prescription, preparation, administration, clinical and laboratory control and final evaluation (Mascarenhas, 2015).

Ministry of Health and Health Surveillance Ordinance No. 272 (1998) standardises the environmental and structural requirements necessary for handling NP. All industrially prepared correlatives and pharmaceutical supplies purchased for the preparation of NP must be registered with the Ministry of Health and accompanied by the certificate of analysis issued by the manufacturer, meeting microbiological and physicochemical specifications (Mendes, Medeiros, & Paulo, 2008).

In order for these extemporaneous formulations to have the quality required in the prescription and after handling, validation of aseptic handling and the environment is necessary, as is technical validation of each handler before they are released for routine handling. Validation is the documented act that aims to ensure that the procedure, process, equipment or operation actually leads to the expected results (RDC n.17, 2010). NP formulations must be handled in a clean area classified as grade A or unidirectional flow, surrounded by a clean area classified as grade B (Portaria MS/SVS n. 272, 1998; RDC n.17, 2010).

As you can see, the quality of the prepared NP formulation is directly related to rigorous quality control, validations, the existence of standard operating procedures, recording the stages of the handling process and, above all, management, vigilance and consistency on the part of both the staff responsible and all employees directly or indirectly involved in the handling of NP (Brazilian Society of Parenteral and Enteral Nutrition & Brazilian Association of Nutrology, 2011).

The cleaning and disinfection of areas, facilities and equipment must be described as operational procedures and these must be available to the responsible and operational personnel. The products selected for cleaning and disinfection must not contaminate facilities or handling equipment with toxic, chemical, volatile or corrosive substances. In addition, disinfectants and detergents must be monitored for microbial contamination (Portaria MS/SVS n. 272, 1998).

However, the whole procedure of preparing NP is complex and critical due to its formulation and the multidisciplinary involvement of nutritional therapy. These peculiarities make the whole process susceptible to errors, which when they reach patients cause various complications and risks to their health. Therefore, although NP is an important therapy in cases of malnutrition in neonates, for example, it requires several precautions to ensure its safety as it is considered a high-risk therapy (Villafranca, Sánchez, Guindo, & Felipe, 2014).

The focus on detecting errors in NP prescriptions is also of great importance in pharmaceutical care, since it is a different kind of manipulation and prescription errors are not limited only to drugs and drug EAs in a hospital environment. As has been pointed out, NP prescriptions are commonly composed basically of carbohydrates, amino acids, lipids, vitamins, electrolytes and minerals, and are a solution or emulsion packaged in a glass or plastic container, which must be sterile and apyrogenic (Ansel & Stoklosa, 2008).

As it is dedicated to the patient, all NP is drawn up using an individualised prescription. The NP is part of a process that involves various professionals such as doctors, pharmacists, nutritionists and nurses, and is extremely complex. It is considered to be one of the main sources of unexpected errors in the treatment of hospitalised patients and is one of the stages in a complex process (Pazin-Filho *et al.*, 2013).

The prescription is forwarded to a *Compounding Center*, which is a compounding pharmacy specialising in the preparation of personalised medicines for patients, where they are prepared on the basis of a prescription prescribed by medical professionals, the individual ingredients of which are mixed together in the form of an exact dosage. This personalised method allows the compounding pharmacist to work with a drug that meets the patient's specific needs (*Professional Compounding Centers of America*, 2015). All the care taken with NP is because it is a preparation that is administered to patients before the results of microbiological analyses, which explains the need to control the working environment and aseptic preparation techniques (Bertol, Adams, & Werlang, 2006; Ordinance of the Ministry of Health and Secretariat of Sanitary Surveillance [MS/SVS] n. 272, 1998).

The medical prescription is considered one of the main sources of unexpected errors in the treatment of hospitalised patients, and is one of the stages in a complex process. This whole process involves a multidisciplinary team: doctors, pharmacists, nurses and nutritionists, each with their own responsibilities and duties (Pazin-Filho *et al.*, 2013). Since the prescription is the initial stage of the care process, with the doctor responsible for its execution, it also participates in or influences the finalisation of another essentially medical process, which is the collection of data to draw up a diagnostic hypothesis and determine a therapeutic plan (Neves & Pazin, 2008). Also according to Pazin-Filho and collaborators (2013), since the prescription is the transition from therapeutic planning to the support team's execution of the patient's treatment, it is seen as the first stage in a series of events within the medication procedure, which will result in the safe or unsafe administration to the patient. The prescription is essentially a communication tool between the doctor, pharmacist, nurse, carer and patient (Néri *et al.*, 2011).

It is a systematised activity in interrelated stages, which provide important elements for a correct prescription. It is therefore essential that prescriptions have the minimum elements needed to achieve an efficient care process (Néri, Viana, & Campos, 2008). In this case, all prescribing professionals need to be aware of the need for their prescriptions to be clear, objective and complete, minimising the doubts of all professionals and providing favourable conditions for drug therapy (Gimenes *et al.*, 2010).

The greater the complexity of the process, the greater the occurrence of errors, so strategies must be devised to avoid the possibility of this turning into harm to the patient, since errors are an inherent part of human learning (Pazin-Filho *et al.*, 2013). The large number of drugs and commercial products available on the market, the high frequency of new launches and the huge number of interactions and adverse effects produced by these medicines make this important stage of the care process susceptible to errors (Wen, 2000). Cassiani, Freire and Gimenes (2003) also discussed the increase in the number of medicines available on the market, as well as the various routes of administration and complex therapeutic regimes as factors related to medication errors, due to the difficulties health professionals have in constantly updating their knowledge, anticipating the indications and side effects characteristic of each medicine.

Drug-related problems are common and are classified as the main cause of adverse events, most of which are considered avoidable and constitute medication errors (Bedouch *et al.*, 2012). These errors are responsible for a significant increase in length of stay, morbidity, mortality and hospital costs, which can be effectively prevented through clinical pharmaceutical interventions (Lucca, Ramesh, Narahari, & Minaz, 2012; Khalili, Farsaei, Rezaee, & Khavidaki, 2011).

Medication errors are any preventable event that can cause harm to the patient when medicines are under the responsibility of healthcare professionals and can occur throughout the medication process (Silva & Cassiani, 2004). They affect patients in all age groups, with a higher incidence and severity in paediatric patients due to their physiological and behavioural characteristics at this stage of development (Sard *et al.*, 2008). According to the study by Kaushal *et al.* (2001), medication errors with the potential to cause harm are three times more common in paediatric inpatients than in adults (Figure 1).

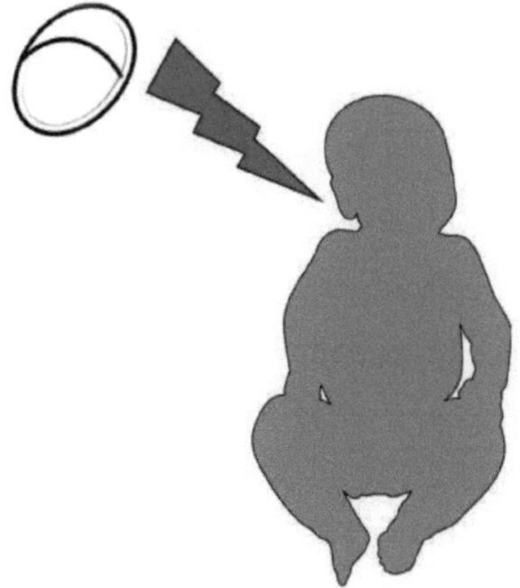

Figure 1. Paediatric patients are more vulnerable to medication errors.
Source: Prepared by the authors.

The paediatric population is subject to medication errors due to the large variation in body mass that requires the calculation of single drug doses based on the weight or body surface area, age and clinical condition of the patient (Ghaleb *et al.*, 2006). A high rate of medication errors in children is found in the prescription and administration stages during the medication process, according to the results of systematic reviews and original studies (Kaushal *et al.,* 2001; Cowley, Williams, & Cousins, 2001; Fortescue *et al.,* 2003; Miller, Robinson, Lubomski, Rinke, & Pronovost, 2007). Koumpagioti *et al.* (2014) evaluated 18 original articles providing an overview of the literature review on paediatric prescription errors. Among the articles, nine mentioned prescription errors exclusively, five pointed out prescription and administration errors, one mentioned prescription and release errors and three of them pointed out all types of errors.

According to Cassiani (2000), there are important definitions for understanding this process: adverse events (or reactions) to medication are defined as any injury due to medication. These events can be preventable, such as a wrong dose, or non-preventable, such as redness due to an antibiotic. Medication error is defined as any error that occurred during the medication process, in the prescription, dispensing, administration and monitoring phases. Potential adverse events are events in which a medication error occurred but did not cause injury for some reason (for example, the error was intercepted before the patient was affected or the patient received the wrong dose but no injury occurred). All potential events are medication errors, but not all medication errors are potential events.

The literature indicates that errors are unlikely to have a single cause, such as professionals, and should be considered more as a failure of the system than a failure of individuals (Pepper, 1995). Research has therefore been carried out analysing the drug distribution system, methods and instruments for detecting and evaluating errors, in an attempt to shift the focus away from causes directed exclusively at professionals (Dean, Barber & Barker, 1995).

Communication failures are often the main causes of errors and, consequently, adverse events (Silva, Cassiani, Miasso, & Opitz, 2007). The former are defined as the failure to carry out a planned action or the application of an incorrect plan. The latter are defined as incidents that result in harm to a patient; for example, if a unit of blood was infused incorrectly and the patient died of a haemolytic reaction (Runciman *et al.*, 2009). These complications affect an average of 10 per cent of hospital admissions and are one of the biggest challenges for improving quality in healthcare (Gallotti, 2004).

In a study carried out by Jacobsen, Mussi and Silveira (2015) on the analysis of prescription errors in a hospital in the southern region of Brazil, 2687 prescriptions were analysed and the following errors were detected:

incomplete dosage (92.7%), absence of pharmaceutical form (83.1%), presence of abbreviations (70.3%), absence of age (63.7%), absence of hospitalisation unit (57.1%), absence of concentration (38.2%), absence of professional's stamp (17.6%), absence of bed (16.3%), presence of code, number (15.8%), illegibility (13.2%), absence of professional council registration number (12.6%), presence of erasures (12.2%), incomplete patient name (7.9%), absence of date (2.0%), absence of administration route (1.3%) and absence of professional signature (0.9%). Medication errors are a serious problem in healthcare services, and are one of the main adverse events suffered by hospitalised patients. Dosage errors are probably one of the most frequent of all medication errors in hospitals (Joanna Briggs Institute [JBI], 2007).

Error reporting is of unique importance because it makes it possible to analyse the factor that caused the error. However, there are estimates that only 5% of medication errors are reported and the main factor behind this underreporting is fear of punishment (Cassiani, 2000). A punitive culture still prevails in situations where an error or adverse event occurs, as evidenced by reports of reprimanding and punishing practices by members of the nursing team. In this context, communication barriers lead to greater adverse events, as information about the consequences of adverse events and the proposal of more effective alternatives for managing them is not passed on, preventing a satisfactory resolution (Leitão *et al.*, 2013).

On the road to successfully creating a non-punitive environment that encourages the reporting of errors, it must be made apparent that errors are caused by problems in the system, not by negligent professionals. Creating a non-punitive system thus means creating a culture that accepts the fact that individuals make mistakes and developing a system that makes it impossible for individuals to make mistakes. Computerised reporting can be, among several alternatives, a way of increasing the capture of incidents (Cassiani, 2000).

From this perspective, it is of fundamental importance that healthcare organisations establish specific policies and procedures for reporting and disclosing adverse events. According to RDC no. 36 of 25 July 2013, the management of the health service must set up the Patient Safety Centre (NSP) which, among its duties, is responsible for notifying the National Health Surveillance System (SNVS) of adverse events arising from the provision of health services.

Considering the need to develop strategies, products and actions aimed at health managers, professionals and users on patient safety, which would make it possible to promote the mitigation of the occurrence of adverse events in health care, the National Patient Safety Programme (PNSP) was instituted on 1° April 2013 by means of Ministry of Health and Minister's Office Ordinance No. 529, which establishes the national patient safety programme. Its aim is to contribute to the qualification of healthcare in all healthcare establishments throughout the country (Leitão *et al.*, 2013; Ordinance MS/GM n. 529, 2013).

It is well known that prescriptions play an important role in the prevention of adverse events, especially those related to incorrect dosage, because when they are ambiguous, incomplete, illegible or the nomenclature of prescribed medicines is not standardised, as well as the use of abbreviations or erasures, they can contribute to the presence of errors (Cassiani & Gimenes, 2003). According to the Guide to Good Practices in Hospital Pharmacy and Health Services drawn up by the Brazilian Society of Hospital Pharmacy (SBFH), medical prescriptions should be analysed by the pharmacist for their components, quantity, quality, compatibility, interactions, possibility of adverse reactions and stability, among other relevant aspects (Novaes *et al.*, 2009).

According to Lewis *et al.* (2009), an analysis estimated the incidence of prescription errors at a median of 7% of drug requests, 52 errors per 100 admissions and 24 errors per 1000 patients per day. The main adverse event suffered by hospitalised patients is prescription errors, which are considered a serious problem in today's health services (Gimenes *et al.*, 2010). These errors have serious consequences for the patient, in many cases leading to death, as well as consequences and penalties for health professionals and institutions (Sousa, Sousa, Moreira, Lisboa & Lira, 2015).

According to Gimenes *et al.* (2010), who carried out a study in five Brazilian hospitals on medical prescriptions, the highest incidence found in prescriptions was the presence of acronyms and/or abbreviations, which were present in 96.3% of prescriptions. The patient's registration number was also absent from 54.4 per cent of prescriptions and the drug's dosage was absent from 18.1 per cent. Errors involving medicines occur frequently in hospitals and are classified as preventable adverse events, which may or may not result in harm to patients (Camerini & Silva, 2011).

In the hospital environment, patient safety has generated debate worldwide and has received various interpretations, including the idea that safety consists of reducing the risk and unnecessary harm associated with health care to an acceptable minimum. Among the resources available, the use of medication is one of the most widely used, but adverse events and medication-related errors are frequent in the hospital environment (Vincent, 2010).

Surveys carried out in American hospitals in 1994 estimated the costs of morbidity and mortality associated with the use of medicines at around 136 billion dollars a year. Adverse drug reactions were classified as the fourth leading cause of this type of death (Wen, 2000). The traditional process of prescribing medicines is manual and has some limitations that encourage its computerisation, such as: a) the large number of types of medicines; b) the filing and handling of a large number of documents; and c) the low quality of manuscripts (Bell *et al.,* 2005). Prescriptions are of great importance in preventing AEs related to the wrong dose, because when they are incomplete or illegible, as well as in cases where the nomenclature of prescribed medicines is not standardised or when abbreviations and erasures are used, they contribute to the occurrence of errors, according to Gimenes *et al.* (2010). Due to the disturbing situation regarding prescription errors, actions aimed at improving the problem and treating them have emerged as preventative measures. These measures are very important and relevant, given that the number of unacceptable errors is increasing (Abreu, 2013). The risks of medication errors increase like prescription errors, as logical but incorrect assumptions about the medication to be used can be made due to missing information (Gimenes, Miasso, Lyra, & Grou, 2006).

As well as modernisation and technological advances, industrialisation has brought an appreciation of science as opposed to man and his values. In the area of health, there have also been a number of technological advances with the emergence of information technology and various devices that have brought benefits and speed in the fight against diseases. This technology contributes greatly to solving problems that were previously unsolvable and which can lead to better living conditions and health for the patient (Barra *et al.*, 2006).

Information technology (IT) is increasingly being used to support the health of the population, as well as in public health activities promoting surveillance and monitoring, health prevention and promotion, and disease control (Pinochet, 2011).

Information technology *software* in the health sector should contribute to and help efficiency, effectiveness and quality improvement, enabling evidence and the teaching process as well as research. Technology is becoming increasingly necessary, efficient and easily accessible. In recent years, computer processing has doubled every year, reducing its cost by 50 per cent (Marin, 2010).

The essence of the profession is information and, just like any other professional, healthcare professionals need fast and efficient information to carry out their care, management and evaluation activities. Healthcare activities are related to the use of information. In this case, the better the information, the better the professional's decision-making will be, since computerised systems are able to record, store and make all this information available (Marin, 2010). The growing number of new techniques in health care procedures makes this sector one of the most dynamic for absorbing new technologies, with a value that is independent of their effectiveness. They are commodities added to technical procedures, but much more from the perspective of interests than needs. Nowadays, new techniques and more modern devices are appearing on the market all the time, in line with the increasingly rapid technological transformations in the health sector (Barra *et al.*, 2006).

Computer systems must be used to reduce the occurrence of errors, since human error is unexpected. Actions are carried out correctly by relying on these information systems, which are dependent machines that can be controlled and predicted. Systems developed for this purpose are expected to increase the quality of human activities (Marin, 2010).

The IT revolution and the advent of computers have brought the acquisition of information and knowledge to today's society, causing notable changes in their habits and lifestyle (Dalri & Carvalho, 2002). Computer systems must be used to reduce the occurrence of errors, since human error is unexpected. It is easier to carry out actions correctly with these information systems, which are dependent machines that can be controlled and predicted. Systems developed for this purpose are expected to increase the quality of human activities (Marin, 2010). The information age has not left the health sector by the wayside. In fact, technology has gone beyond standard data processing for common administrative functions in all organisations, such as human resources, payroll, accounting systems, among others, and now plays a fundamental role in patient care, in the interpretation of the electrocardiogram, as well as in work schedules, prescriptions, results reporting and prevention systems (Pinochet, 2011).

Information technology is increasingly being used to support the health of the population, as well as in public health activities promoting surveillance and monitoring, health prevention and promotion and disease control (Pinochet, 2011). IT *software* in the health sector should contribute to and aid efficiency, effectiveness and quality improvement, enabling evidence and the teaching process as well as research (Marin, 2010).

Technological evolution in healthcare began with the industrial revolution, through the development of new technologies in all areas of knowledge. At this time, applied sciences made it possible to create machines and equipment that replaced and/or minimised the need for physical human strength. The combination of technology and equipment led to this first industrial revolution (Barra, Nascimento, Martins, Albuquerque, & Erdmann, 2006). As well as modernisation and technological advances, industrialisation brought an appreciation of science as opposed to man and his values. In the area of health, there have also been several technological advances with the emergence of information technology and various devices that have brought benefits and speed in the fight against diseases. This technology contributes greatly to solving problems that were previously unsolvable and which can result in better living conditions and health for the patient (Barra *et al.*, 2006).

The introduction of information technology in the health sector represents a major step forward in consolidating new trends, as it makes it easier to carry out work that is notoriously bureaucratic and increases direct patient care activities (Alves, Moura, & Lopes, 2005).

Technological resources such as computers can provide direct and indirect support in the health area, such as hospital information systems and electronic patient record management systems, providing information that can contribute to decision-making (Marin, 2010). In short, many scientific and technological advances have been achieved thanks to the use of computers, which make it possible to handle massive amounts of information in an organised and rapid manner (Marin, 2006).

As for direct support, the use of computerised systems designed to help in the preparation of diagnoses, treatments and prognostic profiles stands out. Known as decision support systems, these act as support in the decision-making process and are systems developed for healthcare professionals (Marin, 2010). As an example given by Alves *et al.* (2005), a Decision Support System (DSS) is defined as a programme designed to help professionals make clinical decisions, i.e. a computer system that relates data and knowledge to provide support to healthcare professionals.

Other examples of health information systems are nursing, pharmacy and nutrition systems, as well as accounting systems that promote the efficiency of health processes. Also making a contribution to healthcare are billing and budget forecasting systems, which are solutions for maximising the benefits of the services provided, promoting greater efficiency. As well as telemedicine, imaging, expert or decision support systems, which are examples of solutions that help to achieve greater effectiveness, promoting best practices to improve care (Marin, 2010).

An important example of a healthcare information system is the Electronic Patient Record (EPP), which according to Lourenção and Ferreira (2016) describes and records the patient's entire chain of events, from their inclusion in the system, the medical services provided, procedures, prescriptions and tests carried out. It aims to bring together different types of information relating to the patient's state of health and the care provided to them, representing a fundamental source of information in the decision-making process.

According to Thaines *et al.* (2009), data and information are commonly considered to be the same thing. However, there is an important difference between these concepts and it is important to clarify their peculiarities, especially when it comes to health data and information, as they are the drivers of action policies in this sector. The concept of data is configured as a sequence of quantifiable symbols that bring the reality in question in numerical form and which, on their own, do not express this reality. For data to make sense and express something, it needs to be interpreted and analysed, in other words, generating information. Just like any other professional, healthcare professionals need fast and efficient information to be able to carry out their care, management and evaluation activities. The use of information is related to all health activities. In this case, the better the information, the better the professional's decision-making will be, since computerised systems are able to record, store and make all this information available (Marin, 2010).

The growing number of new techniques in health care procedures makes this sector one of the most dynamic for absorbing new technologies, with a value that is independent of their effectiveness. They are commodities added to technical procedures, but much more from the perspective of interests than needs. Nowadays, new techniques and more modern devices are appearing on the market all the time, in line with the increasingly rapid technological transformations in the health sector (Barra *et al.*, 2006).

Information management in hospitals and related areas is an essential component in the process of providing patient care. The problem of information management has been made even more difficult by an exponential increase in the amount of data to be managed, the number of professionals controlling the processes and the demands for real-time access. The cost of handling information in hospitals has also been the main factor behind the use of computers in an attempt to provide more data at a lower cost (Pinochet, 2011).

In this scenario, information technology management in hospital organisations plays a decisive role, since information technology is the focus of the greatest innovations, permeates the entire organisation and goes beyond, establishing a new relationship dynamic with all the participants in this market (Pinochet, 2011).

Many healthcare organisations, such as hospitals, laboratories, health plan operators and others, are looking for *software* packages *for their businesses with the* aim of enabling their companies to automate and integrate most of their business processes, share common practices and data throughout the company and produce and access information in real time. This type of integrated management system is called ERP (Enterprise Resource Planning) and is basically characterised by integrating different areas of the organisation into a single application, i.e. a single system with a business process vision, and no longer the vision with the separation of departments that preceded it (Pinochet, 2011).

The government and medical councils have a strong influence on electronic prescription systems. In Brazil, the Federal Council of Medicine [CFM], through resolution 1.639/2002, established a set of technical standards for systems for storing and handling electronic records, with the aim of ensuring the privacy and reliability of the information. Similarly, the Brazilian Society of Health Informatics [SBIS] drew up the Manual of Security, Content and Functionality Requirements for Electronic Health Record Systems [RES], which serves as a parameter for obtaining the various levels of certification for these systems and, consequently, for their legal validity (Brazilian Society of Health Informatics [SBIS], 2004). Similarly, aspects related to technological infrastructure, such as server capacity to support the high demand for simultaneous processes, adequate bandwidth sizing for data traffic and privacy and security, are also fundamental for electronic prescriptions (Horan *et al.*, 2005; Tulu, 2005). E-prescription systems hold out the promise of increased efficiency and reduced costs. On the other hand, the costs of process redesign, software development, system integration and infrastructure adaptation can be high, negatively influencing these initiatives (Mundy & Chadwick, 2004).

There are interoperability standards called HL7, *Health Level Seven,* which refers to the seventh layer of the ISO/OSI model of network protocols that have been the most widely used for the functional, semantic and operational exchange of data in the health area for more than 20 years on the international scene. The HL7 Brazil Institute was created in 2007 with the aim of providing and promoting standards related to the exchange, integration, sharing and retrieval of electronic information to support medical and administrative practice, allowing for greater control of health services (Sociedade Brasileira de Informática em Saúde [SBIS], 2016).

Electronic prescribing: challenges in adhering to computerised systems

Based on the great technological advances, it can be deduced that both people and organisations in the health sector will benefit from great evolution and transformation. Reality, however, may prove to be contrary to this deduction. The expected evolution and transformation will come to those organisations that have the ability to competently and efficiently manage their technologies and their external and internal life cycles, thus guaranteeing competitive advantage and market conquest (Pinochet, 2011).

One development as a result of the digital age is the use of electronic prescriptions, which, among many functions, aim to unify the phases of the medication supply procedure, thus reducing errors in each phase and consequently prescription errors (Pazin-Filho *et al.*, 2013). Electronic prescriptions are those in which the healthcare professional, instead of manually writing the prescription on a prescription or model form, uses *software* following an established pattern. This method has the capacity to reduce errors as it eliminates the difficulty in reading and understanding caused by the prescriber's illegible handwriting, making it possible for possible errors in its preparation to be corrected at the same time without there being any erasures making it difficult to understand the information (Shane, 2002 as cited in Cassiani,

Freire, & Gimenes, 2003 p.52).

According to Bates *et al.* (1999), electronic prescribing is the process in which the doctor writes the prescription directly on the computer and sends it electronically to the pharmacy, avoiding errors due to lack of understanding of illegible letters or ambiguous and incomplete prescriptions. According to the author, electronic prescribing can reduce medication errors by up to 80 per cent. The electronic prescription system helps doctors draw up prescriptions with the aim of reducing medication errors, especially those related to writing the prescription (Fialho, Torres, & Silva, 2011). In this system, the computer is able to suggest alternatives to the prescribed medication and a diagnostic study, according to information from the clinical doctor, helping to structure the prescription, checking allergies, interactions with other medications, as well as the frequency with which the medication is being given to the patient (Cassiani, 2000).

The main advantage of electronic prescriptions is their legibility, as well as the fact that they contain much more complete information. This eradicates the illegibility of handwriting and reduces related errors, promoting greater safety in the dispensing, preparation and administration of prescribed medicines (Barker *et al.*, 2002). Kalmeijer et al. (2003 as cited in Gimenes 2006 p.16) also point to more legible and complete prescriptions as advantages, as well as faster, more efficient and greater accessibility to patient data.

Also according to Volpe *et al.*, 2016, abbreviations or acronyms are commonly used to save time and are a risk factor, as they can be misinterpreted by healthcare professionals. As well as illegibility, which increases the risk of medication errors. These risk factors were analysed and improved with the electronic prescription system. On the other hand, other professionals have identified errors in electronic prescriptions, which shows that computerised systems do not completely eradicate the possibility of medication errors. However, advantages such as legibility, practicality, organisation and speed were reported by these professionals (Gimenes, Miasso, Lyra, & Grou, 2006).

Within this context, in electronic prescribing, the medical professional writes the prescriptions on a computerised system and, although there is no standardisation for these systems, in general they allow complete access to drug data, either by trade name or by active ingredient, with the aim of avoiding errors in transcribing drug names and drug interactions. Each prescription automatically triggers the necessary processes for the related sectors, such as pharmacy, nursing and billing (Joia & Magalhães, 2009). One factor associated with errors in manual prescriptions is the time spent clarifying faults and, in this case, computerised systems offer a solution through *software* that prevents users from drawing up prescriptions without all the necessary information (Gimenes, Miasso, Lyra, & Grou, 2006). According to Pazin-Filho *et al.* (2013), there is evidence that this reduction exists, but on the other hand there are claims that this resource can interfere with students' and residents' learning about prescribing. However, such a claim is complicated by the fact that medical curricula do not pay enough attention to this teaching or leave it to practical training, which is often insufficient to achieve this goal. What should be done is continuous training in this skill, regardless of the type of resource that is used, because changes occur all the time and we shouldn't go against them, but rather prepare ourselves as much as possible to adapt.

Research has shown that professional performance and medication error rates can improve with the use of electronic tools or computerised systems, supporting pharmacotherapeutic decision-making which often includes checking for drug interactions, although there is still not much evidence of improved patient outcomes (Field *et al.*, 2009). The use of technology favours immediate care, more accurate diagnosis and provides more security for the entire multidisciplinary team of professionals, but it can also contribute to the process of dehumanisation, making the relationship between patient and professional colder and more distant, making the patient feel abandoned, insignificant and invisible, being just part of a cog in the wheel (Barra *et al.*, 2006). In many countries, due to their criticality and the possibility of errors, some health information systems go through a process of quality certification and basic principles of security and confidentiality (Pinochet, 2011). The act of validating contributes in a unique way to the continuous improvement of the processes adopted in the development of products directly linked to human health (Araújo & Brunier, 2012).

In the pharmaceutical industry, as in other health sectors, critical processes must, by law, use the validation of computerised systems required by the agency that regulates the sector (ANVISA). A computerised system can be characterised as a process because it receives inputs (data), processes them and transforms them into information. When these processes are managed, there is a greater chance of achieving quality. The validation stage is important for meeting legal demands and the overall performance of industries seeking to guarantee the reliability of the process being controlled (Araújo & Brunier, 2012). The importance of validating computerised systems is directly related to the impact on patient health, product quality and the integrity of the data that must prove that the system did exactly what it was supposed to do (Araújo & Brunier, 2012). Validation consists of establishing documented evidence, with a high level of certainty, that a specific process will perform effectively and consistently produce a result that meets its previously determined specifications and characteristics (Resolução da Diretoria Colegiada [RDC] n.17, 2010).

The Computerised Systems Validation Guide is a guide drawn up by ANVISA to help implement RDC 17, 2010 by managing and validating computerised systems that have an impact on Good Manufacturing Practices. The Guide describes the activities and responsibilities related to the validation of computerised systems, optimising the activities involved (Araújo & Brunier, 2012). Validating an electronic prescription system would require staff time, which in a given hospital environment is scarce. A study conducted by Joia and Magalhães (2009) pointed out that some doctors who used the electronic prescription system asked the medical directorate about the authenticity and legal validity of the electronic prescription system. In order to respond to the doctors, the implementation team investigated the legal aspects related to electronic medical records and discovered that even if the doctor's photo and signature were printed via the system, the prescription would not be a legally valid document.

The implementation team concluded that a series of security conditions in the infrastructure and data storage would need to be created and strictly followed in order for the system to comply with the standards of the Brazilian Secretariat for Health Informatics [SBIS], the government organisation responsible for regulating and validating systems of this type. The team also discovered that the SBIS requires medical record systems to create an individual digital certificate for users, as well as installing biometric devices on the computers that access the system. After analysing the investments needed to adapt the system to SBIS standards, the implementation team opted to print the prescriptions on paper, adding the doctor's stamp and manual signature, in addition to the one already printed by the system (Joia & Magalhães, 2009).

The cost of validating and legalising electronic prescriptions in Brazilian hospitals doesn't seem to be the only barrier, given the advantages of implementing them. According to Devine *et al.* (2010), although it represents a better readability of electronic prescription systems, reductions in prescription errors, implementation of drug interaction alerts and hypersensitivities, these advantages are not directly linked to their cost.

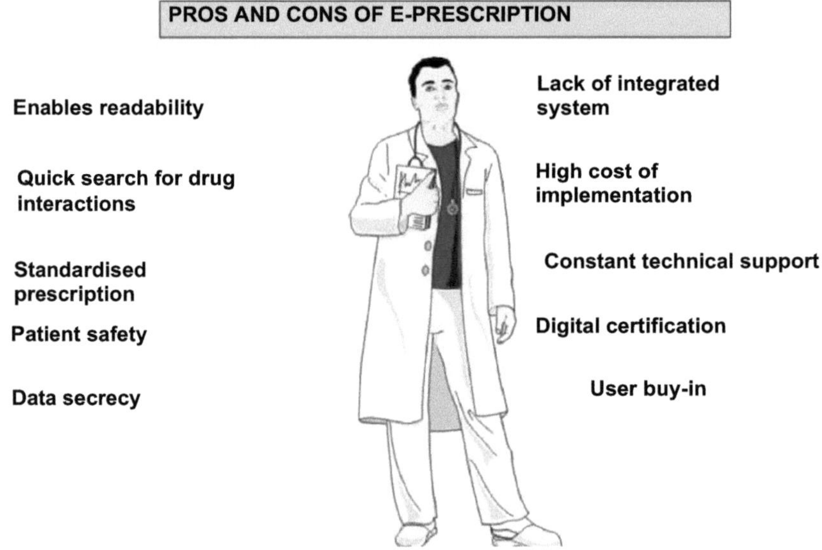

Figure 2: Pros and cons of electronic prescribing.

Source: Prepared by the authors.

However, research has shown that many of the aforementioned initiatives have not achieved their objectives. In many cases, the doctor's resistance to using computers is recognised as the main reason for the underuse or failure of initiatives of this nature (Horan, Tulu, & Hilton, 2005; Lapointe & Rivard, 2005). Electronic prescriptions are not error-prone, although they are a major advance among the strategies designed to minimise the risks of medication errors. In this case, making doctors and residents aware of the importance of training programmes for using the electronic system is extremely important, in order to limit adverse errors as a result of poorly prepared prescriptions (Cassiani, Freire, & Gimenes, 2003). With the implementation and use of these systems and greater training for prescribing professionals, it will be possible to make prescriptions more complete and understandable for the professionals who use them (Gimenes, Miasso, Lyra, & Grou, 2006).

Some obstacles noted in many institutions to these changes and the implementation of an electronic prescription system include hospitals' lack of experience with quality improvement methods, the difficulty of getting doctors involved and the difficulty of placing medication safety as a high priority within the hospital. Therefore, in order to build a safe system, it is necessary to standardise the medication system, standardise high-risk drugs and equipment, learn from errors, eliminate transcription, know the patient's medication history and improve efficiency by analysing error reports (Cassiani, 2000). Data presented by the Regional Centre for Studies on the Development of the Information Society (CETIC.br, 2014) showed that the vast majority of healthcare establishments in Brazil use computers (94%) and the internet (91%) in their activities. However, in the workplace, access to computers and the internet by doctors is lower than the availability of these tools in the establishment in general. Among health professionals, 63 per cent of doctors have access to a computer at work and 60 per cent to the internet. With regard to updating opportunities, 23% of doctors said they had taken part in some course or training on the use of ICT in health.Silva and Marques (2011) pointed out that in the globalised world, information is made available in an exacerbated form and, inherent in this, technology is present, making it necessary to use ICT to process this information as quickly as possible. According to Joia and Magalhães (2009) it is important for the system to be perceived by its potential users and the lack of technological education serves as a barrier to the process of technological implementation, since the age group of the doctor also proved to be an important factor in the acceptance and use of the system. Professionals with more years of experience in medical practice, because they are not used to using computers in their day-to-day work, need careful instrumental and conceptual training in order to take proper advantage of electronic prescriptions.For Markus (1983), there are three alternative vectors that generate resistance to information systems. The first vector involves personal factors. Examples of this vector are, for example: lack of training, fear of computers and the user's perceived lack of utility in relation to the system. The second vector assumes issues related to system *design*. Systems lacking in flexibility, with a graphic interface and usability perceived as weak, overly complex and inadequately designed tend to be rejected or under-utilised by users. For the third vector, the interaction of characteristics related to the system and characteristics related to the organisational context. Examples of this interaction vector could be: systems that centralise data control are resisted in organisations with decentralised authority structures, or systems that balance the distribution of power in organisations will be resisted by those who hold it.Soon, with the help of artificial intelligence technologies together with clinical databases, healthcare professionals will be able to make decisions based on a patient's situation, selecting, for example, the best antibiotic, thus saving costs on medicines and reducing mortality (Pinochet, 2011). Although computerised prescription systems represent a major advance in minimising errors due to illegible and poorly formulated prescriptions, some modifications are necessary in order to improve existing systems. The adverse risks suffered by patients can be reduced with the success of these programmes, improving the quality of care (Gimenes, Miasso, Lyra, & Grou, 2006).

Due to the many potential errors related to prescriptions, the importance of the pharmacist in the analysis prior to the distribution of medicines is evident, minimising possible harm to patients. Therefore, through pharmaceutical intervention, it is possible to reduce adverse events, increase quality of care, reduce hospital costs and promote the rational use of medicines (Jacobsen, Mussi & Silveira, 2015).

Araújo and Uchôa (2011) pointed out that one of the strategies to minimise prescription errors is the implementation of electronic prescriptions, training prescribers, making them aware of and educating them about the importance of correct and legible prescriptions, and the work of clinical pharmacists in the hospital setting. According to the literature, the electronic prescription system has enabled better quality of care for hospitalised patients and others involved, demonstrating that opting for this model can contribute to reducing errors (Nuckols, Smith-Spangler, Morton, Asch, Patel, Anderson, Deichsel & Shekelle, 2014).

In several countries around the world, the topic of electronic prescriptions has been the subject of research. Research has shown that following the implementation of electronic prescribing, there has been a reduction in the frequency of medication errors (Nuckols *et al.*, 2014; Hug, Wilkowski, Sox, Keohane, Seger, Yoon, Matheny & Bates, 2009). There are international studies that have investigated and compared electronic and manual prescriptions (Menendez, Alonso, Rancaño, Corte, Herranz & Vazquez, 2012). These studies showed that after implementing an electronic prescription system, there was a reduction in prescription error rates and an improvement in patient outcomes (Volpe *et al.*, 2016).

Greater investment in infrastructure, training for future medical professionals in information systems, validation with guaranteed security of patient data are still the points to be elucidated if the standardisation of electronic prescriptions is to become a reality and the technology is to be fully implemented in the public and private systems in Brazil. Until then, multi-professional interaction is still the best way to guarantee patient safety and minimise the risk of adverse effects when receiving medication or parenteral nutrition during patient care.

REFERENCES

Abreu, F. G. S. (2013). *Medication errors: evaluation of the prescription and perception of nursing professionals.* Course Conclusion Paper, University of Brasília, Ceilândia College, Ceilândia, DF, Brazil.

Alves, V. M., Moura, Z. A., & Lopes, M. V. D. O. (2005). Health information systems: an analysis of their use by nurses. *Revista da Rede de Enfermagem do Nordeste - RevRene, 6*(3), 31-38.

Ansel, H. C., & Stoklosa, J. M. (2008). *Pharmaceutical calculations* (12ª ed.). Porto Alegre: Artmed.

Araújo, P. T. B., & Uchôa, S. A. C. (2011). Evaluation of the quality of drug prescriptions in a teaching hospital. *Ciência e saúde coletiva, 16*(1):1107-1114.

Barker, K. N., Flynn, E. A., Pepper, G. A., Bates, D. W., & Mikeal, R. L. (2002). Medication errors observed in 36 health care facilities. *Archives of internal medicine, 162*(16), 1897-1903. Retrieved April 29, 2016,
from
http://archinte.jamanetwork.com/article.
aspx?articleid=212740&result
click=1#PARTICIPANTSANDMETHODS.

Barra, D. C. C., Nascimento, E. R. P., Martins, J. J., Albuquerque, G. L., & Erdmann, A. L. (2006). Historical evolution and impact of technology in health and nursing. *Revista Eletrônica de Enfermagem, 8*(3), 422-430.

Bates, D. W., Teich, J. M., Lee, J., Seger, D., Kuperman, G. J., Ma'Luf, N., Boyle, D., & Leape, L. (1999). The impact of computerised physician order entry on medication error prevention. *Journal of the American Medical Informatics Association, 6*(4), 313-321.

Bedouch, P., Tessier, A., Baudrant, M., Labarere, J., Foroni, L., *et al.* (2012). Computerised physician order entry system combined with on-ward pharmacist: analysis of pharmacists' interventions. *Journal of Evaluation in Clinical Practice, 18*(4), 911-8.

Bertol, C., Adams, A. I. H., & Werlang, M. C. (2006). Procedure for validation of the parenteral nutrition preparation process. *Brazilian Journal of Clinical Nutrition, 21*(3), 193-197.

Bottoni, A., Hassan, D. Z., Nacarato, A., Garnes, S. A., & Bottoni, A. (2014). Why worry about hospital malnutrition: literature review. *Journal of the Health Sciences Institute, 32*(3), 314-7.

Braúna, C. C., & Freitas, R. M. (2014). Integrative Review on the use of computerised systems in the practice of pharmaceutical care. *Revista Eletrônica de Farmácia, XI*(1), 35-47.

Camerini, F. G., & Silva, L. D. da S. (2011). Patient safety: Analysis of intravenous medication preparation in a sentinel network hospital. *Texto Contexto Enfermagem, Florianópolis 20*(1), 41-9.

Carvalho, A. P. P. F., Modesto, A. C. F., Oliveira, C. P., Penhavel, F. A. S., Vaz, I. M. F, *et al.* (2014). *Enteral and parenteral nutritional therapy protocol of the nutritional support committee.* Goiânia: Hospital das Clínicas da Universidade Federal de Goiás.

Cassiani, S. H. B., Freire, C. C., & Gimenes, F. R. E. (2003). Electronic medical prescriptions in a university hospital: writing faults and users' opinions. *Revista da Escola de Enfermagem da USP, 37*(4), 51-60.

Cassiani, S. H. B. (2000). Medication errors: prevention strategies. *Revista Brasileira de Enfermagem. 53*(3), 424-430.

Regional Centre for Studies on the Development of the Information Society (2014). Internet Sector Panorama. ICT in the health sector: availability and use of information and communication technologies in Brazilian health establishments. *Technology and Health, 6*(1), 1-10.

Cowley, E., Williams, R., & Cousins, D. (2001). Medication errors in children: a descriptive summary of medication error reports submitted to the United States pharmacopeia. *Current Therapeutic Research, 62*: 627-640.

Dalri, M. C. B., & Carvalho, E. C. (2002). Planning nursing care for patients with burns using *software*: Application to four patients. *Revista Latino-Americana de Enfermagem, 10*(6), 787-793.

Dean, B. S. E. L., Barber, N. D., Barker, K. N. (1995). Comparison of medication errors in an American and British hospital. *American Journal of Health System Pharmacy, 52*(22), 2543-2549.

Devine, E. B., Williams, E. C., Martin, D. P., Sittig, D. F., Tarczy- Hornoch, P., *et al.* (2010). Prescriber and staff perceptions of an electronic prescribing system in primary care: a qualitative assessment. *BMC medical informatics and decision making, 10*(1), 1.

Dudziak, E. A., Gabriel, M. A., Villela, M. C. O. (2000). The education of library users in the face of the knowledge society and its

insertion in the new educational paradigms. Retrieved 04, Nov. 2016,
from
https://www.researchgate.net/profile/Elisabeth_Dudziak/publication/4

H05708 The education of university library users before the knowledge_society and their insertion into the new educational paradigms/links/0deec52f36178be3e9000000.pdf.

Ferreira, I. K. C. (2007). Nutritional therapy in the intensive care unit. *Revista Brasileira de Terapia Intensiva, 19*(1), 90-97.

Fialho, R. C. N., Torres, N., Jr. & Silva, C. M. V. A. (2011) The use of *poka-yoke* to reduce failures in health services: A case study of the medical prescription process in a clinic specialising in the prevention and treatment of neoplastic diseases. *Anais SIMPOI - Simpósio de Administração da Produção, Logística e Operações Internacionais, São Paulo, SP.*

Field, T. S., Rochon, P., Lee, M., Gavendo, L., Baril, J. L., *et al.* (2009). Computerised clinical decision support during medication ordering for long-term care residents with renal insufficiency. *Journal of the American Medical Informatics Association, 16*(4), 480-485.

Fortescue, E. B., Kaushal, R., Landrigan, C. P., McKenna, K. J., Clapp, M. D., *et al.* (2003). Prioritising strategies for preventing medication errors and adverse drug events in paediatric inpatients. *Pediatrics, 111:*722-729.

Gallotti, R. M. D. (2004). Adverse events: what are they? *Journal of the Brazilian Medical Association, 50*(2), 114-114.

Garófolo, A., Boin, S. G., Modesto, P. C., & Petrilli, A. S. (2007). Evaluation of the efficiency of parenteral nutrition regarding energy supply in paediatric oncology patients. *Revista de Nutrição, 20*(2), 181-190.

Ghaleb, M. A., Barber, N., Franklin, B. D., Yeung, V. W., Khaki, Z. F., Wong, I. C. (2006). Systematic review of medication errors in paediatric patients. *Annals of Pharmacotherapy, 40*:1766-1776.

Gimenes, F. R. E., Mota, M. L. S., Teixeira, T. C. A., Silva, A. E. B. C., Opitz, S. P., *et al.* (2010). Patient safety in drug therapy and the influence of medical prescription on dose errors. *Revista Latino-Americana de Enfermagem, 18*(6), 1055-1061.

Horan, T., Tulu, B., & Hilton, B. (2005). Understanding physician use of online systems: an empirical assessment of an electronic disability evaluation system. In R. W. Schuring & T. A. M. Spil (Eds.). *E-health systems diffusion and use: the innovation, the user and the use IT model.* Hershey, PA, USA: Idea Group Inc.

Hug, B. L., Witkowski, D. J., Sox, C. M., Keohane, C. A., Seger, D. L., *et al.* (2009). Adverse Drug Event Rates in Six Community Hospitals and the Potential Impact of Computerised Physician Order Entry for

Prevention. *Journal of General Internal Medicine, 25*(1), 31-38. Retrieved November 15, 2016, from http://doi.org/10.1007/s11606-009-1141-3.

Jacobsen, T. F., Mussi, M. M., & Silveira, M. P. T. (2015). Analysis of prescription errors in a hospital in the southern region of Brazil. *Brazilian journal of hospital pharmacy and health services, 6*(3), 23-26.

Joanna Briggs Institute (2007). Strategies to reduce medication errors with reference to older adults. *Management in Health, 11*(1).

Joia, L.A, & Magalhães, C. (2009). Empirical evidence of resistance to the implementation of electronic prescribing: an exploratory-exploratory analysis. *RAC-Eletrônica, 3* (1), 81-104.

Khalili, H., Farsaei, S., Rezaee, H., & Dashti-Khavidaki, S. (2011). Role of clinical pharmacists' interventions in detection and prevention of medication errors in a medical ward. *International Journal of Clinical Pharmacy, 33*, 281-284.

Kaushal, R., Bates, D. W., Landrigan, C., McKenna, K. J., Clapp, M. D., *et al.* (2001). Medication errors and adverse drug events in paediatric inpatients. *Journal of the American Medical Association - JAMA, 285*:2114-2120.

Koumpagioti, D., Varounis, C., Kletsiou, E., Nteli, C., & Matziou, V. (2014). Evaluation of the medication process in paediatric patients: a meta-analysis. *Journal of Paediatrics, 90*: 344-355.

Lapointe, L., & Rivard, S. (2005). A multilevel model of resistance to information technology implementation. *MIS Quarterly, 29*(3), 461491.

Law no. 5.991, of 17 December 1973 (1973). Provides for the sanitary control of trade in drugs, medicines, pharmaceutical supplies and related products, and makes other provisions. Retrieved 29May , 2016, from http://www.planalto.gov.br/ccivil_03/leis/L5991.htm

Leitão, I. M. T. D. A., Oliveira, R. M., Leite, S. D. S., Sobral, M. C., Figueiredo, S. V., *et al.* (2014). Analysis of adverse event reporting from the perspective of care nurses. *Revista da Rede de Enfermagem do Nordeste - RevRene, 14*(6) 1073-83.

Lewis, P. J., Dornan, T., Taylor, D., Tully, M. P., Wass, V., *et al.* (2009). Prevalence, incidence and nature of prescribing errors in hospital inpatients. *Drug safety*, 32(5), 379-389.

Lopes, L. N., Garcia, K. P., Dias, L. G., Soares, L. R., Leite, A. M., *et al.* (2014). Quality of medical prescriptions in a School Health Centre in the Brazilian Amazon. *Journal of* the *Brazilian Society* of *Clinical Medicine, 12*(2).

Lourenção, L. G., & Ferreira, C. D. J., Jr. (2016). Implementation of the electronic patient record in Brazil. *Enfermagem Brasil, 15*(1), 44-53.

Lucca, J. M., Ramesh, M., Narahari, G. M., & Minaz, N. (2012). Impact of clinical pharmacist interventions on the cost of drug therapy in intensive care units of a tertiary care teaching hospital. *Journal of Pharmacology and Pharmacotherapeutics, 3*(3), 242-247.

Marin, H. F. (2006). Current Perspectives on Nursing Informatics. *Revista Brasileira de Enfermagem, 59*(3): 354-357.

Marin, H. F. (2010). Health information systems: general considerations. *Journal of Health Informatics, 2*(1), 20-4.

Markus, L. M. (1983). Power, politics, and MIS implementation. *Communications of the ACM, 26*(6), 430-444.

Martins, G. A., & Theóphilo, C. R. (2009). *Scientific Research Methodology for Applied Social Sciences.* São Paulo: Atlas.

Mascarenhas, M. B. J., Barros, R. S., Martins, B. C. C., Loureiro, C. V., Araújo, T.D. de V., *et al.* (2015). Neonatal parenteral nutrition solutions in a Brazilian teaching hospital: from indication to administration. *Brazilian Journal of Hospital Pharmacy and Health Services, 6*(2), 18-23.

Mendes, A. S., Medeiros, P. A. D., & Paulo, P. T. C. (2008). Profile of patients undergoing parenteral nutrition therapy in a public hospital. *Revista Brasileira de Farmácia, 89*(4), 373-375.

Mendonça, L., & Silva, A. M. L. (2015). Evaluation of quality control in adult and paediatric parenteral nutrition carried out in a specialist nutrition clinic. *Caderno de Graduação Ciências Biológicas e da Saúde, 2*(3), 25-37.

Menendez, M. D., Alonso, J., Rancaño, I., Corte, J. J., Herranz, V., & Vazquez, F. (2012). Impact of computerised physician order entry on medication errors. *Revista De Calidad Asistencial, 27*(6), 334-340.

Miller, M. R., Robinson, K. A., Lubomski, L. H., Rinke, M. L., & Pronovost, P. J. (2007). Medication errors in paediatric care: a systematic review of epidemiology and an evaluation of evidence supporting reduction strategy recommendations. *Quality & safety in health care, 16*:116-126.

Néri, E. D. R., Gadêlha, P. G. C., Maia, S. G., da Silva Pereira, A. G., de Almeida, P. C., &et *al.* (2011). Medication prescription errors in a Brazilian hospital. *Journal of the Brazilian Medical Association, 57*(3), 306-314.

Néri, E. D. R., Viana, P. R., & Campos, T. A. (2008). Tips for a good hospital prescription. Fortaleza: Federal University of Ceará. Walter Cantídio University Hospital.

Neves, F. F., & Pazin-Filho, A. (2008). Clinical reasoning in the emergency room. *Medicina (Ribeirão Preto. Online), 41*(3) 339-346.

Novaes, M. R. C. G., Souza, N. N. R., Néri, E. D. R., Carvalho, F. D., Bernardino, H. M. O. M., & Marcos, J. F. (2009). *Guide to Good Practices in Hospital Pharmacy and Health Services.* São Paulo: Ateliê Vide o Verso, 356p.

Nuckols, T. K., Smith-Spangler, C., Morton, S. C., Asch, S. M., Patel,V. M., *et al.* (2014). The effectiveness of computerised order entry at reducing preventable adverse drug events and medication errors in hospital settings: a systematic review and meta-analysis. *Systematic Reviews,* 3(56) 1-12. Retrieved November 15, 2016, from: http://doi.org/10.1186/2046-4053-3-56.

Pazin-Filho, A., Frezza, G., Matsuno, A. K., de Alcântara, S. T., Cassiolato, S., *et al.* (2013). Principles of hospital prescribing for medical students. *Medicina (Ribeirão Preto. Online), 46*(2), 183-194.

Pepper, G.A. (1995). Errors in drug administration by nurses. *American Journal of Health System Pharmacy, 52* (4), p. 390-395.

Pinochet, L.H.C. (2011). Information technology trends in health management. *Mundo Saúde, 35*(4), 382-394.

Portaria MS/GM n. 529, de 1^0 April 2013 (2013). Retrieved 29 April 2016 from http://bvsms.saude.gov.br/bvs/saudelegis/gm/2013/prt0529_01_04_2013.html.

Ordinance MS/SNVS n. 272, of 8 April 1998 (1998). Retrieved 27 March 2016 from http://portal.anvisa.gov.br/wps/wcm/connect/d5fa69004745761c8411d43fbc4c6735/PORTARIA 272 1988.pdf?MOD=AJPERES.

Professional Compounding Centres of America, Inc.What Is PCCA? Retrieved March 30, 2016, from: http://www.pccarx.com/what-is-compounding/what-is-compounding.

Collegiate Board Resolution no. 17, of 16 April 2010 (2010). Provides for Good Manufacturing Practices for Medicines. Retrieved 16 April, 2016, from http://bvsms.saude.gov.br/bvs/saudelegis/anvisa/2010/res0017_16_04 _2010.html

Collegiate Board Resolution no. 60, of 6 July 2000 (2000). Establishes the minimum requirements for Enteral Nutrition Therapy. Retrieved 16 April, 2016 from http://portal.anvisa.gov.br/wps/wcm/connect/61e1d380474597399f7b df3fbc4c6735/RCD+N°+63-2000.pdf?MOD=AJPERES

Collegiate Board Resolution no. 36 of 25 July 2013 (2013). Institutes actions for patient safety in health services and makes other provisions. Retrieved 1° May, 2016, from http://bvsms.saude.gov.br/bvs/saudelegis/anvisa/2013/rdc0036 25 07 2013.html.

Rocamora, J. A. I. (2001). Long-term total parenteral nutrition with growth, development and positive nitrogen balance. Surgery 1968. 64:134- 142. *Hospital Nutrition, XVI*(6) 286-292.

Rodrigues, J. M. F., & Sobreira, M. J. (2013). Pharmaceutical monitoring of a patient using parenteral nutrition. *Brazilian Journal of Hospital* Pharmacy and *Health Services, 4*(2), 24-27.

Runciman, W., Hibbert, P., Thomson, R., Van Der Schaaf, T., Sherman, H., *et al.* (2009). Towards an International Classification for Patient Safety: key concepts and terms. *International Journal for Quality in Health Care, 21*(1), 18-26.

Silva, A. E. B. C., & Cassiani, S. H. B. (2004). Medication administration: a systemic vision for the development of preventive measures for medication errors. *Revista Eletrônica de Enfermagem, 6*(2), 279-285.

Silva, A. E. B. C., Cassiani, S. H. B., Miasso, A. I., & Opitz, S. P. (2007). Communication problems: a possible cause of medication errors. *Acta Paulista* de *Enfermagem, 20*(3), 272-6.

Silva, I. S. A. da & Marques, I. R. (2011). Knowledge and barriers in the use of Information and Communication Technology resources by nursing teachers. *Journal of Health Informatics, 3(1): 3-8.*

Brazilian Society of Parenteral and Enteral Nutrition & Brazilian Association of Nutrology (2011). *Recommendations for Parenteral Nutrition Preparation.* Retrieved 16 April, 2016, from http://www.projetodiretrizes.org.br/9 volume/recommendation-for-parenteral-nutrition-preparation eparo da nutricao parenteral.pdf.

Brazilian Society of Health Informatics (2016). *Certification Manual for Electronic Record Systems in* Retrieved 7 September 2016, from http://www.sbis.org.br/certificacao/Manuahttp://www.sbis.org.br/certi ficacao/Manual Certification SBIS- CFM

Sousa, J. B. G., Sousa, B. G., Moreira, A. R., Lisboa, A. R., & Lira, D. L. F. (2015). Pharmaceutical analysis of medical prescriptions in the intensive care unit (ICU) of the Cajazeiras-PB regional hospital. *FAMA Journal of Health Sciences, 1*(2), 1-10.

Thaines, G. H. L. S., Bellato, R., Faria, A. P. S., Araújo, L. F. S. (2009). Production, flow and analysis of health information system data: an exemplary case. *Texto e Contexto de Enfermagem, 18*(3): 466-474.

Vincent, C. (2010). *Patient Safety: Guidelines for preventing adverse events* (1ª ed.). Porto Alegre: Yendis.

Villafranca, J. J. A., Sánchez, A. G., Guindo, M. N., & Felipe, V. F. (2014). Using failure mode and effects analysis to improve the safety of neonatal parenteral nutrition. *American Society of Health-System Pharmacists, 71*: 1210-1218.

Volpe, C. R. G., Melo, E. M. M, Aguiar, L. B., Pinho, D. L. M., Stival, M. M. (2016). Risk factors for medication errors in electronic and manual prescribing. *Latin American Journal of Nursing, 24*: e2742.

Waitzberg, D. L. (2002). *Oral, enteral and parenteral nutrition in clinical practice* (3ª ed.). São Paulo: Atheneu.

Yokoyama, C. S. (2010). *Information system for pharmacotherapeutic monitoring in a community pharmacy.* Master's thesis, Pontifical Catholic University of Paraná, Curitiba, Paraná, Brazil.

Table of contents

CHAPTER 1 1
CHAPTER 2 4
CHAPTER 3 11
CHAPTER 4 19
CHAPTER 5 34
REFERENCES 36

Buy your books fast and straightforward online - at one of world's fastest growing online book stores! Environmentally sound due to Print-on-Demand technologies.

Buy your books online at
www.morebooks.shop

Kaufen Sie Ihre Bücher schnell und unkompliziert online – auf einer der am schnellsten wachsenden Buchhandelsplattformen weltweit! Dank Print-On-Demand umwelt- und ressourcenschonend produzi ert.

Bücher schneller online kaufen
www.morebooks.shop

Printed by Books on Demand GmbH, Norderstedt / Germany